Table of Contents

1 Introduction to Wind Energy

Harnessing the power of the wind represents a significant step towards a sustainable energy future. Wind energy, unlike fossil fuels, produces no greenhouse gas emissions during operation, making it a vital tool in mitigating climate change. This clean energy source offers a compelling alternative to traditional power generation, presenting a path towards energy independence and reduced reliance on finite resources. The potential is immense; consider the sheer volume of kinetic energy constantly swirling around us, largely untapped until relatively recently.

The global wind energy landscape is one of impressive growth, though unevenly distributed. Europe, particularly countries like Denmark and Germany, have been pioneers in wind power adoption, boasting significant installed capacity and well-established infrastructure. Similarly, China has made substantial investments in wind energy, rapidly expanding its wind farms both onshore and offshore. While these regions represent a large portion of global wind power generation, the potential for expansion extends to numerous other areas, particularly in regions with consistent strong winds and supportive policies. Developing nations can leverage wind energy to overcome energy poverty and foster economic growth.

Understanding the fundamentals of wind turbine operation is crucial. These complex machines convert the kinetic energy of moving air into rotational energy, which is then transformed into electricity. The efficiency of this conversion process is paramount, determining the overall cost-effectiveness and environmental impact of wind energy generation. This efficiency is heavily influenced by several factors, including turbine design, blade material, and the sophisticated control systems that manage energy production and grid integration. The interplay of these components dictates the overall performance and economic viability of a wind energy project.

The design and engineering of wind turbine blades are particularly critical. The shape, size, and material composition of these blades directly influence their ability to capture wind energy. Modern blade designs often incorporate advanced aerodynamic principles and computational fluid dynamics to optimize energy capture. The materials used in blade construction, such as fiberglass reinforced polymers and carbon fiber composites, must be durable, lightweight, and able to withstand the significant stresses encountered during operation. Continuous research and innovation are key to enhancing the efficiency and longevity of these vital components.

Efficient energy conversion and transmission are equally important in maximizing the output of wind turbines. Modern generators utilize sophisticated technologies to convert rotational energy into electricity with minimal loss. Power electronics play a crucial role in controlling the flow of electricity, ensuring stable output and seamless integration into the power grid. Efficient transmission systems minimize energy losses during the transport of electricity from the wind turbine to the substation. These systems often use high-voltage transmission lines to reduce energy loss over long distances. Careful attention to these factors allows for the extraction of maximum energy from wind and its efficient delivery to consumers.

Beyond the technical aspects of wind energy, economic factors and supportive policies are critical to its successful integration into the global energy mix. Cost-effective wind turbine designs and efficient manufacturing processes are essential for ensuring the competitiveness of wind energy. Government policies, including subsidies, tax incentives, and renewable energy standards, significantly influence the rate of wind power adoption. The increasing competitiveness of wind energy, spurred by technological advancements and supportive policy frameworks, is driving its growth and integration into many parts of the world. The long-term economic viability of wind energy is becoming increasingly apparent, making it a sound investment for both public and private sectors. Furthermore, evolving market trends and increasing consumer demand for renewable energy further propel the wind energy sector forward, fostering innovation and ensuring its continued prominence in the global energy landscape.

1.1 The Promise of Wind Power

Harnessing the wind's power offers a compelling solution to our global energy needs. Its clean, abundant nature presents a compelling alternative to fossil fuels, reducing our reliance on finite resources and mitigating the devastating effects of climate change. The potential is vast, and the rewards—a cleaner, more sustainable future—are worth pursuing with vigor and innovation.

The economic benefits are equally significant. Wind power projects create jobs in manufacturing, construction, and maintenance, stimulating local economies and reducing our dependence on volatile global energy markets. Furthermore, the long-term cost savings associated with reduced reliance on fossil fuels are substantial, offering both individual and societal financial advantages. These projects, however, necessitate careful planning and management to ensure maximum efficiency and long-term sustainability.

However, the integration of wind power into existing energy grids poses considerable challenges. The intermittent nature of wind necessitates sophisticated energy storage solutions and smart grid technologies to manage fluctuations in energy supply. This requires careful coordination between wind farms, transmission networks, and the overall electricity grid. Innovative solutions such as advanced battery storage systems, pumped hydro storage, and demand-side management strategies are crucial for effectively integrating this clean energy source. Moreover, the environmental impact of wind farms must be carefully considered. While wind power itself is environmentally benign, its implementation can affect local ecosystems. Careful site selection, minimizing habitat disruption, and implementing measures to protect wildlife are crucial aspects of responsible wind energy development. Environmental impact assessments are essential, ensuring the long-term health of our planet remains a priority alongside our energy needs. Technological advancements are constantly improving the efficiency and effectiveness of wind power. Developments in turbine design, such as the use of lighter, stronger materials and more efficient blade designs, are increasing energy output and reducing operational costs. Advances in control systems and smart grid integration further enhance the reliability and stability of wind power generation. This continuous drive for efficiency is paramount to the wide-scale adoption of this vital renewable resource.

Beyond technological enhancements, the political and regulatory landscape plays a critical role in shaping the future of wind power. Supportive government policies, such as tax incentives and renewable energy standards, are essential for encouraging investment and accelerating the deployment of wind farms. Stable regulatory frameworks, providing clear guidelines and procedures, are equally vital for attracting investors and ensuring the long-term success of wind power projects. This supportive ecosystem fuels innovation and deployment.

Furthermore, collaboration between stakeholders—governments, businesses, and communities—is crucial for ensuring the successful integration of wind power into society. Open communication, transparent decision-making, and community engagement are essential for building public support and addressing any concerns or misconceptions about wind energy. This cooperative approach is critical for navigating any potential obstacles.

Finally, the future of wind power looks bright. Technological advancements continue to push the boundaries of what's possible, with innovations such as floating offshore wind farms expanding the potential capacity significantly. Hybrid systems, combining wind power with other renewable sources, are also gaining traction, offering diversified and resilient energy solutions. By embracing innovation and collaboration, we can harness the transformative power of wind energy to create a cleaner, more sustainable, and secure energy future for generations to come. The transition may present challenges, but the promise of clean, renewable energy and a healthier planet is a compelling force that should drive progress forward.

1.2 Global Wind Energy Landscape

The global wind energy landscape is a dynamic and rapidly evolving field, characterized by significant regional disparities in both capacity and growth. China, the United States, and several European nations currently lead in installed wind power capacity, reflecting substantial investments in infrastructure and supportive government policies. However, emerging economies in Asia, Africa, and Latin America possess vast untapped wind resources, presenting immense opportunities for future development. This uneven distribution highlights the need for international cooperation and technology transfer to facilitate the global expansion of wind energy. Access to funding and technology remains a critical barrier for many developing nations, emphasizing the importance of targeted financial assistance and capacity-building initiatives.

While onshore wind power remains the dominant technology, offshore wind farms are experiencing exponential growth, particularly in countries with extensive coastlines and favorable ocean conditions. These offshore installations offer higher capacity factors due to consistent and stronger winds compared to onshore sites. However, offshore wind development faces considerable technological and logistical challenges, including higher construction and maintenance costs, as well as the complexities of working in harsh marine environments. The development of more efficient and cost-effective offshore wind turbines is critical to unlocking this vast renewable energy potential. Furthermore, addressing the environmental concerns associated with offshore wind farms, such as potential impacts on marine ecosystems and migratory birds, is crucial for their responsible deployment.

The economic viability of wind energy projects is heavily influenced by factors such as wind resource availability, land or water lease costs, turbine technology selection, and grid connection infrastructure. A comprehensive cost-benefit analysis is essential for developers to assess the financial feasibility of individual projects. Moreover, government policies play a pivotal role in shaping the wind energy market. Supportive policies such as tax incentives, feed-in tariffs, and renewable portfolio standards incentivize investment in wind energy and stimulate market growth. Conversely, inconsistent or unfavorable regulatory frameworks can hinder development by increasing risks and uncertainties for investors. Understanding the interplay between economic factors and policy frameworks is therefore crucial for successful wind energy project development and deployment. Careful consideration of these factors, and a thorough understanding of the local regulatory landscape, are essential for any potential investment or involvement in a project.

Technological advancements are driving down the cost of wind energy and enhancing its efficiency. Improvements in turbine design, including the use of lighter and stronger materials, advanced blade aerodynamics, and optimized control systems, contribute to higher energy capture and reduced operational costs. Likewise, innovations in energy storage technologies, such as battery systems and pumped hydro storage, are addressing the intermittency challenge associated with wind power, allowing for more reliable and predictable energy supply. The ongoing research and development in these areas are essential for sustaining the long-term growth and competitiveness of the wind energy sector. It is worth mentioning the role of data analysis and predictive modeling in optimizing wind farm performance and grid integration.

The integration of wind energy into existing power grids presents both opportunities and challenges. Wind power's variable nature necessitates sophisticated grid management strategies to maintain system stability and reliability. Smart grid technologies, including advanced sensors, communication networks, and real-time control systems, enable better forecasting of wind energy output and optimize grid operations. However, substantial investments in grid infrastructure are often required to accommodate the increasing penetration of variable renewable energy sources. This requires careful planning and coordination between wind energy developers, grid operators, and policymakers to ensure a seamless and efficient integration of wind power into the overall energy system. Understanding the complexity of this integration and the associated costs are vital for successful deployment. Without this understanding, problems with the grid are likely.

Looking towards the future, the continued expansion of wind energy will depend on addressing several key challenges. These include the need for further technological innovation to reduce costs, enhance efficiency, and improve reliability; the importance of effective environmental management to minimize potential impacts on ecosystems; and the necessity for sound policy frameworks to support the long-term growth and sustainability of the wind energy sector. The development of hybrid renewable energy systems, combining wind power with other renewable energy sources such as solar, hydro, and biomass, offers a path towards a more diversified and resilient energy portfolio. Similarly, advancements in energy storage will further enhance the reliability and value of wind energy, enabling its broader penetration into the global energy mix. These are only some of the topics which should be addressed if we want to see a bright future for wind power.

2 Turbine Design & Aerodynamics

The heart of wind energy generation lies in the turbine, a complex machine demanding meticulous design and a deep understanding of aerodynamics. Optimizing blade design is paramount; considerations extend beyond simple shape, delving into airfoil profiles, twist distributions, and the precise placement of structural reinforcements. Computational Fluid Dynamics (CFD) simulations play a crucial role, allowing engineers to virtually test different designs under various wind conditions, identifying areas for improvement and predicting performance characteristics before physical prototypes are even built. This iterative process, combining advanced software with rigorous testing, helps to minimize material waste and maximize efficiency. The goal is not merely to create a functional blade, but one that extracts the maximum possible energy from each gust of wind.

Material selection significantly impacts turbine lifespan and performance. Modern turbines are increasingly incorporating composite materials like carbon fiber and fiberglass, offering a superior strength-to-weight ratio compared to traditional steel. This translates to lighter blades, which require less energy to rotate, leading to higher overall efficiency. These materials also exhibit excellent fatigue resistance, crucial for enduring the constant stress imposed by cyclical loading under varying wind speeds. However, the manufacturing process of these composites requires careful consideration; controlling the curing process, fiber orientation, and resin distribution is key to ensuring consistent quality and preventing premature failure. Further research into novel materials, such as bio-based composites or advanced polymers, promises even greater improvements in the future, potentially leading to lighter, stronger, and more sustainable turbines.

Accurate aerodynamic modeling is essential for predicting turbine behavior and optimizing its design. This involves employing sophisticated software to simulate airflow around the blades, accounting for factors like turbulence, wind shear, and blade interaction. These models enable engineers to predict the forces acting on the blades, calculate power output, and assess potential sources of noise or vibration. The complexity of these simulations necessitates powerful computing resources and advanced algorithms. Advanced techniques, such as Large Eddy Simulation (LES), allow for a more accurate representation of turbulent flow, yielding more reliable predictions. Furthermore, the integration of experimental data from wind tunnel tests or on-site measurements allows for continuous refinement and validation of the aerodynamic models, improving their predictive accuracy over time. This rigorous approach contributes to creating highly efficient turbines capable of harnessing wind energy effectively.

Blade design isn't solely about maximizing energy capture; it's also about minimizing the detrimental effects of fatigue. Repeated stress from wind loads, especially during high-velocity gusts, can cause structural damage over time. This is particularly challenging for larger turbines with longer blades subjected to higher loads. Advanced fatigue analysis techniques are used to predict the lifespan of turbine blades, incorporating factors like material properties, stress concentrations, and environmental conditions. Designing blades with appropriate safety margins helps ensure longevity and minimize the risk of catastrophic failure. Regular inspection and maintenance protocols, combined with advanced structural monitoring, contribute to maintaining the long-term integrity and reliability of wind turbines. Minimizing fatigue is a continuous process, constantly refined through advancements in materials science, design techniques, and monitoring technology.

Beyond blade design, the overall configuration of the turbine significantly impacts performance. Factors such as the nacelle design, the hub diameter, and the tower height all play a role in energy capture and operational efficiency. Nacelle design involves optimizing the aerodynamic shape to minimize drag and ensure smooth airflow around the gearbox and generator. The hub diameter affects the blade sweep area and thus the amount of wind captured. Tower height significantly influences the wind speed experienced by the blades; taller towers often experience higher and more consistent wind speeds, increasing energy production. Careful optimization of these factors requires a holistic approach, considering the complex interactions between various components and the specific site characteristics. Ultimately, a well-designed turbine strikes a delicate balance between maximizing energy capture and minimizing operational costs and environmental impact.

2.1 Blade Design Optimization

Understanding the intricacies of blade design is paramount for maximizing wind turbine efficiency. A poorly designed blade can significantly reduce energy capture, leading to substantial financial losses over the turbine's lifespan. Careful consideration must be given to numerous factors, including airfoil shape, twist distribution, and material selection, all of which interact in complex ways to determine overall performance.

The airfoil shape itself, the cross-sectional profile of the blade, significantly impacts lift and drag. Different airfoil designs excel in varying wind conditions; a blade optimized for high-speed winds might perform poorly in low-speed conditions, and vice-versa. This necessitates detailed computational fluid dynamics (CFD) simulations to evaluate various airfoil options under a broad range of anticipated wind speeds and turbulence levels. These simulations, often leveraging sophisticated software packages, help engineers predict blade performance with impressive accuracy, minimizing the need for expensive and time-consuming physical prototypes. The goal is to achieve a balance between high lift at optimal angles of attack and low drag to reduce energy losses due to friction.

Another crucial design element is the blade's twist distribution. The angle of attack, the angle between the wind direction and the chord line (the line connecting the leading and trailing edges of the airfoil), needs to change along the blade's span. The twist is carefully engineered to ensure that each section of the blade operates at its most efficient angle of attack across a wide range of wind speeds. This is especially important as the wind speed generally increases with altitude, and the blade's tip moves faster than the base. Precise control over the twist distribution allows the blade to harness energy effectively across its entire length, avoiding stall conditions which greatly reduce efficiency and can cause blade damage. This sophisticated design requires advanced computational tools and often iterative refinements based on detailed performance modelling.

The materials used in blade construction are also pivotal in achieving optimal performance and longevity. Modern wind turbine blades often incorporate composite materials, such as fiberglass-reinforced polymers and carbon fiber reinforced polymers. These materials offer a superior strength-to-weight ratio compared to traditional materials, allowing for longer blades that can reach higher wind speeds. A lighter blade reduces stress on the turbine's drivetrain and tower structure, contributing to enhanced reliability and potentially extending the lifespan of the entire wind turbine system. Furthermore, material selection influences factors like blade stiffness, fatigue resistance, and resistance to environmental factors, such as ultraviolet radiation and moisture. Careful analysis of material properties, using advanced material characterization techniques, is essential for creating durable and efficient blades capable of withstanding the harsh conditions of prolonged outdoor exposure.

Beyond these core aspects, blade design optimization also includes considerations such as blade root design, the structural connection between the blade and the hub, and the overall structural integrity of the blade. This involves careful analysis of stress distribution under various operating conditions, incorporating factors such as centrifugal forces, aerodynamic loads, and the potentially harsh environmental conditions. Finite element analysis (FEA) is commonly employed in this process to model the complex stresses experienced by the blade throughout its operating life. This powerful computational method allows engineers to identify potential weak points and optimize the blade's structural design to prevent failures or fractures and ensure safe and reliable operation under all conditions. The results from such FEA simulations feed directly back into design iterations, further refining the blade's geometry and material properties for improved robustness.

Finally, the integration of advanced technologies, such as active flow control mechanisms and novel blade designs inspired by natural systems, such as bird wings, continues to push the boundaries of blade optimization. These advanced designs promise further improvements in aerodynamic performance and energy capture, representing exciting avenues for future research and development in this field. The future of wind energy hinges on continuous innovation and enhancement of turbine blade designs, ensuring that wind energy will continue to play a vital role in the global transition toward sustainable energy sources. This requires a multidisciplinary approach combining engineering expertise, computational modeling, and a deep understanding of the complex interactions between fluid mechanics, material science, and structural mechanics.

2.2 Aerodynamic Modeling

Accurate aerodynamic modeling is crucial for optimizing wind turbine design. Understanding how air flows around the blades is paramount; this dictates the power generated and the overall efficiency of the entire system. We utilize sophisticated computational fluid dynamics (CFD) simulations, which solve complex equations describing fluid motion, to analyze this airflow. These simulations allow us to predict blade performance under various wind conditions, enabling informed design choices. The level of detail in these simulations can range from basic to highly advanced, depending on the specific needs of the project and the resources available.

More advanced models account for complex phenomena like turbulence, blade deformation, and even the impact of atmospheric conditions. These factors significantly affect the accuracy of our predictions. For instance, turbulence in the wind can cause unpredictable variations in lift and drag forces on the blades. Similarly, the flexibility of the blades themselves, influenced by the wind loads, affects the overall aerodynamic performance. Accounting for these complexities enhances the precision of the simulation and ultimately leads to a more efficient turbine design.

The process begins with defining the geometry of the turbine blades, often using detailed CAD models. These models provide the basis for the CFD simulation, defining the surfaces and volumes involved in the airflow analysis. Then, the simulation itself involves setting up the boundary conditions, such as wind speed, direction, and turbulence intensity. This is a crucial step as any inaccuracies here propagate through the entire simulation, leading to potentially misleading results. Once set up, the software iteratively solves the governing equations, generating a comprehensive dataset of velocity, pressure, and force distributions around the blades.

Data visualization is key to interpreting these results. Specialized software tools allow engineers to visualize airflow patterns, pressure contours, and force distributions across the blade surface. This detailed visualization helps identify areas of high stress, low efficiency, or potential design flaws. Identifying these problem areas early in the design process allows for proactive modifications before physical prototypes are built, leading to significant cost and time savings. The iterative process of simulation, analysis, and design modification is essential to optimizing turbine performance.

Beyond CFD, other modeling techniques are used. Blade element momentum (BEM) theory provides a simplified approach, breaking down the blade into smaller segments and analyzing the forces on each segment individually. While less computationally intensive than CFD, BEM lacks the detail and accuracy of CFD, particularly for complex turbine geometries and flow conditions. It's useful for preliminary assessments, however, as it offers a faster means of determining basic performance characteristics. It's especially valuable during the initial stages of design where rapid assessment and iteration are key.

Selecting the appropriate modeling technique depends heavily on the specific goals of the project and the available resources. For detailed design optimization, high-fidelity CFD simulations are essential. For preliminary assessments or when computational resources are limited, BEM theory can be a suitable alternative. Often, a combined approach is adopted, using BEM for initial design and CFD for refinement and detailed analysis. This integrated approach balances the need for speed and accuracy, leveraging the strengths of each modeling method. Understanding the strengths and limitations of each method is crucial for effective aerodynamic design.

Validation of the models is paramount to ensuring their reliability. This often involves comparing model predictions with experimental data, such as wind tunnel tests on scaled-down models or measurements from real-world turbines. Any discrepancies between the model predictions and the experimental data highlight areas requiring further refinement or adjustments in the model parameters. Such refinement and calibration refine the models, thereby enhancing the accuracy of future predictions. The iterative process of model development and validation is crucial for achieving highly accurate and reliable aerodynamic models.

Ultimately, accurate aerodynamic modeling directly translates to improved wind turbine performance, leading to increased energy generation and reduced operational costs. By accurately predicting the complex interplay between blade design, wind conditions, and turbine performance, engineers can optimize turbine design for maximum efficiency, resulting in greater economic return on investment and a significant contribution to sustainable energy production. The continuous improvement of aerodynamic modeling techniques will further enhance the capabilities of wind turbines and their role in the global transition to clean energy.

2.3 Advanced Materials

The relentless pursuit of higher efficiency and longer lifespan in wind turbine blades necessitates the exploration of advanced materials. Traditional materials like fiberglass-reinforced polymers, while functional, often fall short in extreme weather conditions or over extended operational periods. This limitation underscores the critical need for innovation in material science to address these inherent weaknesses and unlock the full potential of wind energy. Carbon fiber-reinforced polymers (CFRPs), for instance, offer significantly enhanced strength-to-weight ratios and fatigue resistance compared to their fiberglass counterparts. Their superior stiffness translates to more efficient blade designs that can capture more energy from the wind, while their durability means less frequent and costly maintenance. However, the high cost of CFRPs currently limits their widespread adoption, making ongoing research into cost-effective manufacturing processes crucial for their future integration.

Moving beyond CFRPs, research is actively investigating other groundbreaking materials. Advanced composites incorporating nanomaterials, such as carbon nanotubes or graphene, promise even greater enhancements in strength, stiffness, and fatigue life. The incorporation of these nanomaterials into polymer matrices can significantly improve the mechanical properties of the composite, leading to lighter, stronger, and more durable blades capable of withstanding the extreme loads and stresses experienced during operation. These advancements also extend to the development of self-healing materials, which can autonomously repair minor damage, preventing catastrophic failure and extending the operational lifespan of the blades. This potential for self-repair drastically reduces maintenance costs and downtime, a significant advantage in the long-term economic viability of wind farms.

The development of these novel materials isn't simply about enhanced performance; it also encompasses environmental considerations. Sustainable manufacturing processes for these advanced materials are paramount. The environmental impact of producing and disposing of materials needs careful consideration. Lifecycle assessments, which meticulously evaluate the environmental footprint of materials from cradle to grave, are crucial for responsible material selection. Minimizing the carbon footprint of material production is essential for achieving the broader sustainability goals of wind energy. Using recycled materials or bio-based polymers, for example, can contribute to lowering the environmental impact, aligning with the industry's overarching commitment to cleaner energy solutions.

Beyond the blade itself, the application of advanced materials extends to other critical components of wind turbines. For example, innovative materials are being explored for the nacelle and tower structures, aiming to reduce their weight while maintaining structural integrity. Lightweighting these components reduces the overall stress on the turbine, potentially increasing its lifespan and decreasing transportation and installation costs. Advanced alloys, such as high-strength steels or titanium alloys, are being investigated for their potential in increasing the resilience and durability of these parts. These materials promise to enhance the structural robustness of the turbine, improving its operational safety and longevity in demanding environments, including harsh weather conditions and offshore deployments. The cost-effectiveness of employing these materials must be balanced against the potential benefits. A detailed cost-benefit analysis should be undertaken before making a final selection.

Furthermore, the quest for optimal performance demands improvements not only in the materials themselves but also in the manufacturing techniques employed. Advanced manufacturing processes, such as additive manufacturing (3D printing), offer significant potential for creating complex and lightweight blade designs that are unattainable using conventional methods. These processes allow for greater design freedom and improved precision, leading to more efficient blades. However, scalability and cost remain major challenges for the widespread adoption of these advanced manufacturing technologies in the wind energy sector. Continued research and development are needed to overcome these hurdles and unlock their full potential. The integration of these advanced manufacturing techniques, with carefully selected advanced materials, paves the way for a new generation of wind turbines, significantly advancing the overall efficiency and sustainability of wind energy technology. The future of wind energy relies on our ability to continuously innovate and explore the possibilities presented by advanced materials. Their impact stretches far beyond simple performance gains; it is fundamental to building a sustainable and resilient energy future.

3 Energy Conversion & Efficiency

Harnessing the kinetic energy of the wind and transforming it into usable electricity is a complex process, demanding careful consideration at every stage. The efficiency of this conversion directly impacts the overall profitability and environmental benefit of wind energy projects. High-efficiency generators are paramount; permanent magnet synchronous generators (PMSGs) currently dominate, boasting high power density and minimal maintenance requirements. However, research continually pushes the boundaries, exploring novel materials and designs to further enhance efficiency and reduce costs. This involves optimizing the magnetic field strength, minimizing resistance within the generator windings, and ensuring effective heat dissipation. The ongoing quest for improved generator technology is crucial for maximizing the return on investment for wind energy projects.

Power electronics play a pivotal role in managing the fluctuating nature of wind energy. Rectifiers transform the alternating current (AC) output of the generator into direct current (DC), which is then inverted back into AC suitable for the grid using sophisticated power electronic converters. These converters not only regulate voltage and frequency but also control the power flow, optimizing energy transfer and minimizing losses. Advancements in switching devices, like Insulated Gate Bipolar Transistors (IGBTs) and Silicon Carbide (SiC) MOSFETs, are leading to higher switching frequencies and reduced energy dissipation, significantly enhancing the overall efficiency. The integration of advanced control algorithms further refines the power electronic systems' performance, dynamically adapting to changing wind conditions and grid demands.

Efficient energy transmission is essential to minimize losses between the wind turbine and the grid. Properly designed transmission lines and transformers are critical; minimizing resistance and reactive power losses is key to reducing transmission losses. High-voltage transmission lines are typically employed to reduce current flow, thereby lowering resistive losses. The careful selection of conductors, insulators, and transformer designs plays a crucial role in system efficiency. Further improvements are actively researched, including the use of superconducting materials for transmission lines which promise near-zero resistance and vastly improved efficiency. The cost-effectiveness of such advanced technologies is however subject to ongoing development and market forces.

Seamless grid integration is paramount for wind energy's widespread adoption. The intermittent nature of wind power requires sophisticated strategies to maintain grid stability and reliability. This involves predictive control algorithms which estimate wind power output and adjust grid operations proactively. Advanced grid management systems monitor wind power fluctuations in real-time, adjusting generation from other sources accordingly. Efficient grid integration relies on reliable communication networks facilitating rapid data exchange between wind farms and the grid control center. The focus on smart grid technologies is essential to manage the increasing integration of renewable energy sources and bolster the stability of the grid.

Minimizing energy losses throughout the entire system is crucial for maximizing overall efficiency. This involves careful attention to detail at every stage, from turbine design to grid connection. Identifying and reducing losses in each component can lead to cumulative improvements in overall energy yield. Regular maintenance and monitoring can detect and correct problems in the early stages, preventing more significant energy losses over time. The use of advanced diagnostic tools helps pinpoint areas of inefficiency, assisting operators in optimizing operations and preventing costly failures. This proactive approach is increasingly important as wind farms become larger and more complex, maximizing returns on investment and minimizing environmental impact.

3.1 Generator Technology

Generators are the heart of a wind turbine, converting the mechanical rotation of the blades into electrical energy. Understanding their technology is crucial for maximizing efficiency and reliability. Different generator types, each with its own strengths and weaknesses, exist. Permanent magnet generators (PMGs) are favored for their simplicity and high efficiency at smaller scales, while doubly-fed induction generators (DFIGs) offer superior control and power regulation capabilities, making them suitable for larger turbines. However, DFIGs have a more complex design and potentially higher maintenance needs. The choice depends on factors like turbine size, grid requirements, and cost considerations. Careful analysis of these factors is essential for optimizing the overall wind energy system.
Selecting the appropriate generator technology involves a multifaceted decision-making process. For instance, the size of the wind turbine directly impacts the choice of generator. Small turbines, often used in residential or remote applications, typically employ PMGs due to their cost-effectiveness and relative simplicity. In contrast, large-scale wind turbines in commercial wind farms generally

favor DFIGs because of their flexibility in controlling reactive power flow and adapting to fluctuating wind conditions, features essential for grid stability at that scale. This choice highlights the vital interplay between turbine scale and generator selection. The grid infrastructure also plays a significant role; a robust grid might support less-flexible generator technologies, whereas less-developed grids might require generators with superior power regulation capabilities.

Beyond PMGs and DFIGs, other generator types exist, though less commonly used. Synchronous generators are a mature technology, offering stable operation and high power quality. However, their efficiency can be slightly lower compared to PMGs or DFIGs. Furthermore, the technology's inherent complexity can translate to increased costs and maintenance requirements. The decision often hinges on a trade-off between initial investment costs and long-term operational efficiency. Other emerging technologies such as superconducting generators are showing promise, particularly in offshore wind farm applications due to their potential for enhanced efficiency and compactness, despite currently facing challenges in terms of cost and cooling requirements. Therefore, continuous evaluation of technological advancements is paramount.

Generator technology is an area of continuous advancement, with ongoing research aimed at enhancing efficiency, reliability, and cost-effectiveness. For instance, improvements in magnet materials have significantly enhanced the power density and efficiency of PMGs. Researchers are constantly exploring new magnet materials with higher energy products, leading to more powerful and smaller generators. This ongoing work highlights the importance of staying updated with the latest developments in the field. Similarly, advances in power electronics, particularly in high-power converters, are improving the controllability and performance of DFIGs. This leads to better power quality and more effective grid integration. Staying informed about these breakthroughs allows for optimization of wind energy systems.

Optimizing generator performance hinges on several factors. Proper cooling is essential to prevent overheating and maintain optimal operating temperatures. Different cooling techniques exist, ranging from air cooling for smaller generators to liquid cooling for large-scale systems. The choice of cooling method depends on the generator size and environmental conditions, but selecting the right one is vital for prolonging the generator's lifespan and ensuring continuous operation. Regular maintenance is also crucial. This includes scheduled inspections, cleaning, and repairs to prevent malfunctions and ensure continuous operation. Preventive maintenance is much cheaper in the long run, than reactive maintenance. Proactive strategies, using condition monitoring techniques like vibration analysis, can identify potential problems before they escalate into major failures, ensuring the maximum possible up-time. A robust maintenance schedule is key to maximizing energy output from the generator.

The integration of generator technology with the overall wind turbine system is critical. Effective system design requires careful consideration of the generator's characteristics, including its torque-speed curve, power output, and control requirements. This integration influences the design of the gearbox, the drive train, and the control system. Proper integration ensures that the generator operates at its optimal efficiency, matching the varying speed and torque output of the wind turbine blades. Furthermore, the generator's control system plays a crucial role in regulating the power output and adapting to changing wind conditions. Advanced control algorithms can enhance efficiency and ensure grid stability. Therefore, the generator should be viewed not as an isolated component but as a fundamental part of a complex, interacting system.

Finally, consider the environmental impact. The materials used in generator manufacturing, their lifecycle, and disposal need careful consideration. Sustainable materials and manufacturing processes help minimize the environmental footprint of the entire wind energy system. Efficient designs minimize energy consumption during operation and maintenance, reducing the environmental impact throughout the entire life cycle of the generator. This should be a consideration throughout the design, selection and operational stages, helping to ensure the sustainability of wind energy production. Therefore, responsible choices in generator technology are essential in delivering genuinely clean and sustainable energy solutions.

3.2 Power Electronics

Power electronics are the backbone of efficient wind energy conversion. They play a crucial role in managing the fluctuating nature of wind power, transforming the variable voltage and frequency output of the wind turbine generator into a consistent and usable form compatible with the electrical grid. This involves sophisticated control systems and power conversion techniques, which are essential for optimizing energy harvesting and ensuring grid stability. Without these advanced electronics, the integration of wind energy into the grid would be significantly more challenging and less efficient.

The heart of this process lies in the power converter, a device that acts as an intermediary between the generator and the grid. Typically, wind turbines use doubly-fed induction generators (DFIGs) or permanent magnet synchronous generators (PMSGs), each requiring a unique power converter configuration optimized for its specific characteristics. DFIGs, for instance, utilize a partial-scale converter connected to the rotor, allowing for independent control of reactive power and active power flow. This precise control contributes to maximizing energy capture and reducing the burden on the grid. PMSGs, on the other hand, generally employ full-scale converters, offering wider control options and often improved efficiency, particularly in smaller wind turbines. The choice between these generator types and associated converter designs depends heavily on the specific application and desired performance trade-offs.

Furthermore, the power electronics systems must handle various voltage levels, protecting both the turbine and the grid. Voltage source converters (VSCs) are commonly employed, providing flexibility in controlling the voltage and frequency of the output, facilitating grid synchronization and seamless power injection. Advanced control algorithms, often employing sophisticated signal processing and predictive techniques, are integrated to ensure smooth operation despite the inherent variability of wind power. These algorithms continuously monitor the wind conditions and generator output, adapting the power electronics' behavior to optimize energy capture while mitigating stress on the system components. Grid codes define stringent requirements for the power quality injected by renewable energy sources. Power electronics play a crucial role in meeting these standards, mitigating harmonic distortion and voltage fluctuations. Filters, both passive and active, are employed to suppress harmonic currents and improve the overall power quality delivered to the grid. Active filtering technologies offer superior performance and adaptability but come at a higher cost. Careful selection of filtering strategies is essential to balance performance requirements with economic considerations.

Moreover, the efficient operation of power electronics is intrinsically linked to the overall efficiency and reliability of the wind energy system. Losses within the converter can significantly detract from the overall energy output, thus necessitating the use of high-efficiency components and careful design optimization. Advanced semiconductor devices such as insulated gate bipolar transistors (IGBTs) and wide bandgap semiconductors, such as silicon carbide (SiC) and gallium nitride (GaN), are increasingly employed to reduce conduction and switching losses, enhancing overall converter efficiency and system reliability. The use of these high-performance semiconductors has been demonstrated to decrease power losses considerably and increase the overall lifespan of the equipment.

Beyond efficiency, power electronics also contribute to the reliability and safety of the wind energy system. Protection circuitry is integrated within the power converters to guard against overcurrents, overvoltages, and faults, preventing damage to both the turbine and the grid. This protection is crucial given the harsh environmental conditions in which wind turbines operate and the inherently unpredictable nature of wind power. Sophisticated monitoring systems continuously track the performance of the power electronics, alerting operators to potential problems and allowing for timely maintenance, improving the overall lifespan and reducing the risk of downtime. These systems must be robust and reliable to ensure the uninterrupted flow of clean energy. The data collected by these monitoring systems can also be instrumental in optimizing the control algorithms over time.

The future of power electronics in wind energy hinges on ongoing advancements in semiconductor technology, control algorithms, and power conversion topologies. The development of more efficient and compact components is a continuous area of research and development, leading to further improvements in energy conversion efficiency and cost reduction. Advanced control techniques, such as model predictive control (MPC) and artificial neural networks (ANNs), offer enhanced control precision and adaptability, further improving the energy harvesting capabilities of wind turbines. The ongoing transition to smarter grids necessitates smarter power converters capable of handling increased levels of variability and integration with other renewable energy sources. The development of more resilient and adaptable power electronics systems will be instrumental in realizing the full potential of wind energy as a sustainable and reliable energy source, ensuring efficient and dependable electricity grids for the future. The continuous improvement in all aspects ensures the long-term viability of renewable wind energy.

3.3 Transmission Systems

Efficient power transmission is paramount in maximizing the output of a wind farm. Losses during transmission can significantly impact the overall energy yield, negating some of the gains achieved through advanced turbine technology. Therefore, understanding and optimizing the transmission system is critical. Minimizing these losses requires careful consideration of several key factors, from cable selection to the implementation of advanced grid technologies.

The choice of transmission cables significantly influences energy loss. High-voltage cables, while more expensive upfront, minimize resistive losses over long distances. The material composition of these cables, the conductor size, and insulation quality all contribute to the overall efficiency. A thorough cost-benefit analysis, considering both the initial investment and long-term operational costs, is essential for making informed decisions. This includes not only the direct cost of the cables but also the potential impact on maintenance schedules and lifespan. A comprehensive life-cycle assessment should be employed.

Beyond the cables themselves, the entire transmission infrastructure plays a crucial role. The design and placement of substations, transformers, and other components must be optimized to reduce power losses. Strategic placement of substations can shorten transmission distances, leading to reduced losses. Similarly, the efficient design of transformers is key; choosing transformers with minimal internal losses can significantly contribute to overall efficiency gains. This requires detailed simulations and modeling to ensure optimal design for varying load conditions and environmental factors such as temperature fluctuations.

Advanced grid technologies offer further opportunities for optimization. Smart grids, equipped with sophisticated monitoring and control systems, allow for real-time adjustments to compensate for fluctuations in wind energy generation. These systems can dynamically adjust power flows across the network, directing energy to areas of high demand while minimizing transmission losses along less-utilized pathways. Moreover, the integration of sophisticated data analytics can assist in predictive maintenance. Early identification and mitigation of potential issues can dramatically increase the reliability and operational efficiency of transmission systems.

Power quality is another critical aspect to address. Fluctuations in wind speed can cause voltage variations, potentially damaging equipment and reducing system stability. Power quality conditioning equipment, such as filters and voltage regulators, is crucial to ensure the stability and reliability of the transmission system and to protect the downstream grid infrastructure. Choosing robust and high-performing conditioning equipment is a crucial investment to avoid costly repairs and potential downtime. This requires evaluating various options and considering long-term maintenance needs. It's also critical to factor in environmental considerations, such as the heat generated by these devices.

Furthermore, the integration of renewable energy sources into the existing grid infrastructure demands careful planning and coordination. Retrofitting existing grids to accommodate high penetrations of wind power often requires substantial upgrades. This might involve reinforcing transmission lines, upgrading substations, and installing new protection equipment. A comprehensive assessment of the existing infrastructure is essential to develop a plan that balances cost-effectiveness with long-term reliability. Efficient upgrades avoid costly and disruptive grid shutdowns. This involves collaboration with grid operators and regulatory bodies to ensure compliance with all relevant standards and regulations.

Finally, the impact of environmental factors on transmission line performance should not be overlooked. Weather conditions, such as high winds and extreme temperatures, can affect the efficiency and stability of the system. Design considerations must account for these factors to ensure reliable operation under diverse climatic conditions. Regular maintenance and inspections are also needed to identify any issues related to environmental wear and tear. It's also important to consider the environmental impact of the transmission infrastructure itself, reducing its footprint on land and minimizing the visual impact on landscapes.

3.4 Grid Integration

Seamless grid integration is paramount for harnessing the full potential of wind energy. It's not merely about connecting turbines to the grid; it's about managing the inherent intermittency of wind power and ensuring grid stability. This requires sophisticated technologies and careful planning to maintain a reliable energy supply.

Effective grid integration begins with accurate forecasting. Predictive models, utilizing historical wind data, meteorological forecasts, and advanced algorithms, can estimate future power output from wind farms with surprising accuracy. This allows grid operators to anticipate fluctuations and adjust accordingly, preventing potential overloads or blackouts. Real-time monitoring systems, equipped with sensors throughout the wind farm and the grid, provide crucial feedback, confirming predictions and identifying any deviations.

The design of the transmission system itself plays a critical role. High-voltage transmission lines are often necessary to minimize energy losses over long distances from remote wind farms. Substations, carefully located to manage the flow of electricity, are essential components. Furthermore, advanced power electronics, such as power converters and grid-forming inverters, play a vital role in smoothing the intermittent power output of wind turbines, converting DC power to AC for grid compatibility and enhancing grid stability. These components need to be robust, reliable, and well-maintained to avoid any interruptions.

Modern grid integration also incorporates smart grid technologies. These technologies facilitate communication and control between the wind farm, the transmission system, and the broader grid. Smart grid capabilities allow for dynamic adjustments based on real-time conditions, ensuring optimal energy flow and managing fluctuations in supply and demand. This sophisticated interplay allows for the most efficient utilization of the energy generated, minimizing waste.

Beyond the technical aspects, regulatory frameworks and market mechanisms are crucial. Clear guidelines for connecting wind farms to the grid, including technical standards and interconnection procedures, are essential. Incentive programs can encourage investments in grid infrastructure upgrades, crucial for accommodating the increasing amount of renewable energy. Additionally, grid operators need to adapt their operations to accommodate the variable nature of wind energy, implementing new scheduling and control strategies. These are essential steps to foster a collaborative and responsive environment.

The cost of grid integration should also be carefully considered. While renewable energy sources offer long-term cost savings, the initial investment in grid upgrades can be substantial. Therefore, a thorough cost-benefit analysis is essential to justify these investments and ensure the economic viability of wind energy projects. This careful assessment helps in determining optimal strategies to minimize overall costs while maximizing long-term benefits.

Finally, grid integration is a continually evolving field. Research and development are critical in developing even more efficient and reliable grid integration solutions. Advanced control algorithms, improved power electronic devices, and more sophisticated forecasting models constantly improve the integration of wind energy into the electrical grid. This ongoing improvement ensures the stability and reliability of the energy supply, making wind energy a truly integral part of the global energy mix. The future of wind power heavily relies on ongoing technological advancements in this specific area.

3.5 Loss Minimization

Minimizing energy losses in wind energy conversion is paramount for maximizing overall system efficiency and economic viability. Every component, from the turbine blades to the grid connection, presents opportunities for optimization and loss reduction. Careful consideration of these areas will significantly improve the return on investment for wind energy projects.

Losses arise from various sources within the system. Mechanical losses occur within the turbine's gearbox and bearings due to friction. These losses can be mitigated through the use of high-efficiency gearboxes employing advanced lubrication techniques and low-friction bearing materials. Regular maintenance, including lubrication schedules and component inspections, plays a critical role in preventing excessive wear and tear, minimizing mechanical loss over the turbine's lifespan. Innovative designs, such as magnetic bearings, completely eliminate the need for lubrication, drastically reducing friction and wear and boosting reliability alongside performance.

Electrical losses occur during the conversion of mechanical energy into electrical energy by the generator. Generator design itself is crucial; using high-efficiency generators with minimized internal resistance significantly improves the conversion process. Furthermore, the use of advanced materials, such as high-temperature superconductors in the generator windings, reduces resistive losses and allows for greater power output with similar designs, improving overall efficiency. Sophisticated control systems can actively monitor and adjust generator operation to optimize performance and further reduce energy loss under various load conditions. The choice of generator technology is critical as well; permanent magnet generators often show better efficiency at certain wind speeds compared to other generator types.

Losses are also present in the power electronics that manage the electricity produced by the wind turbine. This process involves transforming the variable AC power output to constant voltage DC for grid integration. These power conversion stages introduce their own losses, but choosing high-efficiency inverters and optimizing their control algorithms can greatly mitigate this issue. Modern, wide-bandgap semiconductor devices, such as silicon carbide (SiC) and gallium nitride (GaN), exhibit lower switching losses and higher operating frequencies, resulting in improved overall conversion efficiency. Regular monitoring and maintenance of the power electronics are equally important to ensure optimal performance and longevity.

Transmission losses are substantial, especially over long distances from the wind farm to the substation and then to the grid. This is minimized through the use of high-voltage transmission lines and optimized cable design. Careful selection of materials for minimal resistance and minimizing line impedance is vital to this task. The use of advanced grid management systems can also actively route energy to minimize transmission losses. Furthermore, reducing the overall impedance of the transmission lines, with advanced materials or designs, or by simply shortening the distances to substations, will have a positive impact.

Grid integration losses can also be significant. The inherent variability of wind energy necessitates efficient grid integration methods. This includes the use of sophisticated energy storage systems to balance supply and demand, smoothing the output, which would otherwise lead to larger losses and possible power quality issues. Furthermore, the grid itself must be robust and able to handle fluctuating power input without compromising reliability. The development of smarter grid infrastructure equipped with advanced control systems is crucial for minimizing losses during grid integration. This approach allows real-time optimization of energy flow, reducing losses from both fluctuations and transmission issues.

Finally, the complete energy conversion chain must be considered as a system. While minimizing individual component losses is important, optimizing the entire system is even more crucial. This holistic approach involves selecting all components with an eye towards minimizing overall losses, from the turbine design to the grid integration. It involves coordinating the design and operation of each component to work synergistically towards a more efficient overall system.

A well-designed and well-maintained wind energy system will result in significantly reduced energy losses, increased overall efficiency, and ultimately greater economic viability.

4 Smart Grid Integration

Seamless integration of wind energy into the power grid is paramount for its widespread adoption. This necessitates not simply connecting wind turbines, but actively managing their output to maintain grid stability and reliability, a challenge compounded by the inherently intermittent nature of wind. Sophisticated control systems and monitoring technologies are crucial to address this.
Predictive control algorithms, powered by advanced weather forecasting and real-time data analysis, allow for anticipation of fluctuating wind speeds. This proactive approach enables grid operators to adjust power generation from other sources, such as conventional power plants, to compensate for variations in wind energy output, preventing sudden dips or surges in overall supply. These sophisticated algorithms are often based on complex mathematical models that take into account not just wind speed predictions but also factors such as turbine performance characteristics and grid demand forecasts. Effective implementation requires advanced computing infrastructure and robust communication networks.
Real-time monitoring is equally crucial. A network of sensors embedded within the wind farm and at various points along the transmission lines provides a constant stream of data about turbine performance, power output, and grid conditions. This continuous feedback allows for immediate detection of any anomalies, such as turbine malfunctions or grid instability. This data is analyzed by

advanced monitoring systems that use machine learning algorithms for pattern recognition, enabling early detection of potential problems and rapid response to prevent cascading failures. The rapid processing and analysis of this vast dataset require high-performance computing resources and sophisticated software.

The combination of predictive control and real-time monitoring forms a powerful system for managing wind energy integration. It's akin to a skilled conductor orchestrating a complex symphony, ensuring that each instrument (wind turbine, conventional power plant) contributes harmoniously to the overall performance (grid stability). The seamless flow of energy requires continuous communication, coordination, and adaptability.

Moreover, the effectiveness of smart grid integration extends beyond the technical realm. It requires collaboration between wind farm operators, grid managers, and policymakers. Open communication channels and data-sharing agreements ensure efficient coordination and informed decision-making. This collaborative approach addresses concerns regarding grid security and regulatory compliance. Investing in the necessary infrastructure for smart grid integration—high-bandwidth communication networks, advanced computing systems, and sophisticated software—is a substantial but necessary undertaking. However, the long-term benefits far outweigh the initial costs. Reliable and efficient wind energy integration leads to increased grid stability, reduced reliance on fossil fuels, and a significant step toward a sustainable energy future.

Furthermore, the development of standardized communication protocols and data formats is essential for seamless interoperability between different wind farms, grid operators, and control systems. This fosters competition and innovation, leading to improved technology and reduced costs. Collaboration between research institutions, industry partners, and policymakers is critical to defining these standards and ensuring their widespread adoption.

Beyond the immediate benefits of grid stability and reliability, smart grid integration unlocks significant opportunities for optimizing energy markets. By accurately predicting wind energy output, grid operators can better manage energy trading and dispatch, potentially leading to lower energy prices and increased market efficiency. This necessitates the development of sophisticated market models that can account for the variable nature of wind power.

The integration of renewable energy sources into the existing grid infrastructure is a complex undertaking involving a multitude of technical, economic, and regulatory challenges. The design of robust, reliable, and secure smart grids is crucial to maximizing the contribution of wind energy and facilitating the transition to a cleaner and more sustainable energy system. Effective solutions require not only technological advancements, but also careful policy planning, thoughtful regulatory frameworks, and robust collaboration across stakeholders.

Finally, continuous monitoring and adaptation are key. As technology evolves and the integration of renewable energy sources increases, ongoing analysis and refinement of smart grid strategies are necessary to maintain grid stability and reliability. This continuous improvement process requires ongoing investment in research and development, as well as flexible and adaptable regulatory frameworks that can accommodate technological advancements. The dynamic nature of the energy landscape demands a proactive and adaptive approach to smart grid management.

4.1 Predictive Control

Predictive control offers a powerful approach to optimizing wind turbine performance and maximizing energy output. Unlike reactive control systems that simply respond to immediate changes, predictive control algorithms anticipate future wind conditions and adjust turbine operations proactively. This anticipatory approach significantly reduces energy losses caused by fluctuating winds, a key challenge in harnessing wind power effectively. This leads to increased efficiency and more reliable power generation.

The core of predictive control lies in its ability to model the future behavior of the wind. Sophisticated algorithms, often based on advanced forecasting techniques, analyze historical wind data, current wind speeds, and weather forecasts. By extrapolating these data points, the system constructs a predictive model of wind speed and direction over a specific time horizon. This timeframe, typically ranging from seconds to minutes, provides a window for the algorithm to make informed decisions about turbine operation.

Consider a scenario where a sudden gust of wind is predicted. A reactive control system might only react after the gust hits, potentially leading to overspeeding or even damage to the turbine. In contrast, a predictive control system would anticipate the gust. It would adjust the pitch angle of the turbine blades or reduce the generator speed in advance, mitigating the impact of the gust and preventing potential damage or efficiency losses. The anticipatory nature of predictive control is fundamental to its effectiveness.

The implementation of predictive control necessitates the integration of various components. High-quality sensors measuring wind speed and direction are essential for accurate input data. This data forms the basis of the predictive model and drives decision-making within the control system. The model itself is usually computationally intensive, requiring powerful processing capabilities to perform complex calculations in real-time. The control system needs to communicate with the turbine's actuators, such as blade pitch mechanisms and generator speed controllers, to execute the necessary adjustments based on the model's predictions. This complex interplay demands robust communication protocols and reliable data transmission.

Furthermore, model accuracy significantly impacts the efficacy of predictive control. The more precisely the model can predict future wind conditions, the more effective the control strategy will be. To refine model accuracy, advanced techniques like machine learning and artificial neural networks are increasingly being employed. These algorithms can learn from historical data and adapt to varying wind patterns, leading to improved prediction accuracy over time. The incorporation of machine learning adds a layer of sophistication and adaptability to predictive control, allowing it to cope with unpredictable wind behavior more effectively.

A key benefit of this advanced control strategy is its impact on the overall stability and reliability of the power grid. Wind power, by its inherent nature, is an intermittent energy source. Fluctuations in wind speed can cause unpredictable variations in power output, potentially disrupting the grid's balance. Predictive control helps mitigate this intermittency by smoothing out power delivery, resulting in a more stable and predictable energy supply. This improved stability is crucial, particularly as the percentage of wind power in the energy mix increases.

However, designing and implementing a predictive control system is not without its complexities. The development of accurate predictive models requires expertise in wind forecasting, control theory, and data analytics. Moreover, the integration of various hardware and software components necessitates careful planning and execution. The computational demands of the control algorithms can be significant, necessitating high-performance computing resources. Finally, the system must be robust enough to withstand potential failures in its components, ensuring continuous operation even during adverse conditions. Addressing these challenges is paramount to achieving the full potential of predictive control in wind energy.

The future of predictive control in wind energy lies in further refinements and innovations. Advanced machine learning algorithms promise even more accurate wind forecasting. The development of more efficient and powerful computational hardware will enable the real-time processing of increasingly complex models. Finally, the integration of predictive control with other smart grid technologies, such as energy storage and demand-side management, will pave the way for even greater optimization of wind power generation and utilization. Ultimately, the continued refinement and adoption of predictive control will play a crucial role in realizing the full potential of wind energy as a clean and sustainable energy source.

4.2 Real-Time Monitoring

Real-time monitoring of wind turbines is crucial for maximizing energy output and ensuring the reliable operation of the entire wind farm. It allows for proactive identification of potential issues, preventing costly downtime and optimizing energy production. This is achieved through a sophisticated network of sensors strategically placed throughout the turbine, collecting a vast amount of data on various parameters.

This data includes, but is not limited to, blade pitch angles, rotational speed, wind speed and direction at various points, generator temperature, gear box vibrations, and even the condition of the lubricating oil. Sophisticated algorithms then process this data, detecting anomalies and predicting potential failures before they occur. For example, a slight increase in vibration levels might indicate an impending bearing failure, allowing for preventative maintenance to be scheduled before a catastrophic breakdown. This preventative approach minimizes operational disruption and costly repairs.

The immediate benefits extend beyond preventing equipment failure. Real-time data also enables operators to optimize turbine performance in response to changing environmental conditions. By continuously monitoring wind speed and direction, the control system can adjust the blade pitch to maximize energy capture. Similarly, real-time monitoring can identify and address imbalances or inefficiencies within the power conversion system. This ensures energy is converted and transmitted to the grid with maximum efficiency, minimizing energy loss and maximizing profitability.

Moreover, real-time monitoring systems offer valuable insights into the long-term performance of individual turbines and the wind farm as a whole. By analyzing historical data, trends in performance can be identified, highlighting potential areas for improvement or upgrades. This might involve optimizing blade design for specific wind conditions at the particular site, identifying and correcting potential design flaws, or refining operational strategies. The data collected provides a wealth of information for improving the efficiency and profitability of the whole wind farm.

Furthermore, the data gathered from these monitoring systems has implications beyond the immediate operation of the wind farm. This detailed, real-time information can be aggregated and analyzed to improve the design and operation of future wind turbines. This continuous feedback loop from operation to design provides a constant avenue for innovation and improvement within the wind energy industry. This allows for the development of more robust, efficient, and cost-effective wind turbines.

In addition to optimizing energy production and identifying potential issues, real-time monitoring plays a vital role in grid stability. Wind power, being an intermittent source of energy, can pose challenges to grid stability. Real-time monitoring allows for the prediction of power output fluctuations, enabling grid operators to proactively manage these variations and ensure a stable and reliable energy supply. This integration of wind energy into smart grids relies heavily on accurate and timely data. Predictive models built upon real-time monitoring data allow for more precise forecasting of power production, enabling more efficient integration of intermittent renewables into the electricity grid.

Integrating this data with weather forecasting systems further enhances its utility. Combining real-time turbine data with weather predictions allows for more accurate estimations of future energy output, contributing to more effective grid planning and management. This integrated approach minimizes the risks associated with the intermittency of renewable energy sources and further improves grid stability. This sophisticated approach moves beyond reactive management towards proactive and preventative control of the wind farm and its integration into the energy grid.

Finally, the value of real-time monitoring extends to ensuring the safety of personnel and the equipment. Real-time monitoring can detect abnormal operating conditions, such as excessive vibration, high temperatures, or unusual electrical currents, triggering alerts to operators. This allows for timely interventions, preventing potential accidents or damage to the equipment. Early detection of anomalies can prevent catastrophic failures, reduce costly repairs, and minimize the risk of injuries. This combination of efficiency improvements, safety enhancements, and proactive management makes real-time monitoring an indispensable element in the operation of modern wind farms.

5 Wind Farm Siting & Layout

Effective wind farm siting and layout are paramount for maximizing energy production and minimizing environmental impact. A thorough site assessment is the foundation, involving detailed meteorological studies to determine wind speed, direction, and consistency. This includes analyzing long-term wind data from nearby weather stations and employing advanced computational fluid dynamics (CFD) models to simulate wind flow patterns over the terrain. Careful consideration of factors such as wind shear, turbulence intensity, and atmospheric stability is crucial for accurate predictions of turbine performance. Failing to adequately assess these variables can lead to significant underperformance and financial losses. The topography itself plays a significant role; hilly or mountainous regions can create complex airflow patterns that either enhance or disrupt energy capture. Sophisticated modeling is essential to identify locations with the optimal wind resource while considering factors like land availability and accessibility.

Understanding wake effects is essential for optimizing wind farm layout. These occur when the wind passing through a turbine's rotor is slowed and deflected, reducing the efficiency of downwind turbines. This aerodynamic interference can significantly decrease the overall power output of the farm. To mitigate wake effects, turbines must be strategically spaced, taking into account wind direction and turbine characteristics. Numerous techniques exist for wake modelling, ranging from simple empirical equations to more complex CFD simulations. The chosen method depends on the complexity of the site and the desired level of accuracy. In addition to spacing, turbine control strategies can be implemented to actively manage wake effects and improve overall energy capture. For instance, some advanced turbines can adjust their rotor speed and blade pitch to compensate for wind shadowing, although this often involves higher initial investment costs. Careful consideration of both passive and active wake mitigation strategies is needed to optimize efficiency.

Environmental impact assessment forms a crucial part of the siting process. Comprehensive studies must be conducted to evaluate potential effects on wildlife, particularly birds and bats. This may involve using radar systems to monitor bird and bat flight patterns, and ecological assessments to identify sensitive habitats that need protection. Noise pollution is another key consideration, requiring careful modelling and mitigation strategies. This could involve strategic placement of turbines, the use of noise barriers, or operational controls to minimize noise levels during sensitive periods. The aesthetic impact of wind farms on the landscape must also be taken into account, with careful attention paid to visual integration and community engagement. Thorough environmental impact assessments are not only necessary for securing permits but also for building public acceptance of wind energy projects, and avoiding negative publicity which can halt projects. Regulatory frameworks vary from location to location, but a thorough understanding of these regulations is crucial for navigating the permitting process. Working with regulatory bodies early in the planning stage can prevent costly delays and setbacks. Transparency and proactive community engagement throughout the entire process can help manage public perception and foster local support for the project.

Successful wind farm development relies on a synergistic approach that seamlessly integrates meteorological data, advanced wake modelling, and environmental impact assessments. Ignoring these key elements can result in suboptimal energy production, community conflict, and regulatory obstacles. Thorough planning, accurate modeling, and a conscientious approach to environmental considerations are crucial for maximizing the benefits of wind energy while minimizing potential negative impacts. The choice of turbine technology itself impacts both energy production and environmental concerns; larger turbines might produce more energy but also create larger wakes and potential noise issues. Therefore, the selection of turbine type should be integrated into the siting and layout optimization process. This process requires expertise across multiple disciplines – meteorology, engineering, environmental science, and regulatory compliance – to navigate the complexities of wind farm development. The integration of all these factors is not simply a matter of compliance; it's the key to responsible and successful wind energy development that respects both environmental and societal concerns, and provides a path to sustainable energy solutions.

5.1 Site Assessment

A comprehensive site assessment is paramount for successful wind farm development. It's not simply about finding a windy location; it's about meticulously evaluating various factors to ensure optimal energy production, minimal environmental impact, and long-term economic viability. Neglecting this crucial stage can lead to significant financial losses and operational challenges down the line. This means considering everything from the local wind resource to the geological conditions and potential impact on wildlife.

Understanding the wind resource is fundamental. Detailed wind data, gathered over an extended period—ideally a year or more—using anemometers at various heights, is crucial. The analysis of this data provides insights into the average wind speed, turbulence intensity, and wind shear, parameters which are critical for determining the turbine's power output and lifespan. This data forms the backbone of any feasibility study. Without it, estimations become unreliable speculation, risking the success of the project. We must ascertain the consistency and predictability of wind conditions at the proposed location, and how it varies throughout the year. These are not simple values; careful modeling techniques must be applied.

Analyzing the terrain's impact is equally important. Complex terrain can significantly influence wind flow, creating areas of higher or lower wind speeds. Sophisticated computational fluid dynamics (CFD) modeling is often employed to simulate wind patterns, accounting for variations in elevation, vegetation, and nearby structures. This process allows for the optimization of turbine placement to maximize energy capture and to minimize wake effects, where the airflow behind a turbine reduces the performance of those downstream. Proper analysis can avoid costly placement errors that reduce overall farm efficiency. Careful assessment should also identify areas that may be more vulnerable to the erosion or other damage from weather.

Beyond wind and terrain, a thorough site assessment considers environmental factors and regulatory compliance. Environmental impact studies are crucial. These assessments involve evaluating potential impacts on wildlife, including birds and bats. Noise pollution levels must be carefully predicted and mitigated to meet environmental regulations. Furthermore, the site's proximity to existing infrastructure, such as power lines and roads, needs careful consideration to facilitate efficient energy transmission and construction. A detailed understanding of local regulations and permitting procedures is also essential for a smooth project rollout and approval. This also means consulting with relevant authorities and local stakeholders, ensuring the project integrates well with the surrounding ecosystem and community.

Socio-economic factors are also crucial in a holistic assessment. Community acceptance and engagement are essential for a project's long-term success. Consideration of local land use, potential visual impacts on the landscape, and the project's socioeconomic contribution to the surrounding community are vital. Open communication and collaboration with stakeholders can help build support and address concerns proactively. Failure to address community concerns may lead to delays or even the project's cancellation. Therefore, engagement and public relations form an essential aspect of this process. This requires sensitive and proactive communication strategies that are culturally appropriate.

The final component is, of course, economic feasibility. The cost of land acquisition, turbine installation, grid connection, operation and maintenance, as well as any potential environmental remediation, must be carefully analyzed. This analysis helps determine the project's overall profitability and financial viability, factoring in projected energy production and prevailing energy prices. It's equally important to consider potential revenue streams and any government subsidies or tax incentives that might be applicable. A detailed financial model, incorporating various scenarios and uncertainties, is indispensable for informed decision-making. A comprehensive assessment ensures that potential pitfalls are addressed before financial commitments are finalized, reducing the risk of project failure.

5.2 Wake Effects

Understanding wake effects is crucial for optimizing wind farm performance. Turbines placed downwind of others experience reduced wind speed and turbulence, significantly impacting their energy output. This decrease, often referred to as wake loss, can result in a substantial drop in overall farm efficiency, potentially jeopardizing the financial viability of the project. Minimizing these losses is, therefore, a key consideration in wind farm design and layout planning.

Careful site assessment plays a vital role in mitigating wake effects. Sophisticated meteorological data analysis, including wind resource mapping and computational fluid dynamics (CFD) simulations, are employed to predict wind flow patterns and turbulence characteristics across the proposed wind farm site. These analyses help identify optimal turbine placement strategies that minimize wake interference. Factors such as terrain, vegetation, and atmospheric conditions are also carefully considered, adding complexity to the modeling process. Ignoring these factors can lead to significant underperformance, impacting both the energy production and the return on investment. Accurate predictions are essential for successful wind farm projects.

Effective wake modeling involves advanced computational tools. These sophisticated simulations can predict the extent and intensity of wake effects for various turbine arrangements. The models consider various parameters, such as turbine characteristics (rotor diameter, hub height), wind speed and direction, and atmospheric stability. This detailed analysis helps engineers optimize the spacing between turbines and the overall farm layout, maximizing energy production while minimizing wake losses. The higher fidelity of these models is paramount to avoiding significant financial setbacks during project development and construction.

Beyond modeling, innovative strategies exist to reduce wake effects. One promising approach involves controlling individual turbine operation. By adjusting the rotor speed or blade pitch angle, operators can modify the wake characteristics, reducing the impact on downwind turbines. This active wake control requires sophisticated algorithms and real-time monitoring of wind conditions to ensure efficient adjustment and optimization. A practical implication of this is a system which constantly monitors turbine performance and adjusts accordingly, preventing significant losses and maximizing yield.

Another effective technique involves optimizing turbine placement. Strategic arrangement of turbines can help minimize wake overlap. This might involve staggering rows of turbines or varying their spacing to account for variations in wind direction and speed. Understanding the interplay between terrain and wind flow patterns plays a key role here. Careful planning minimizes wind interference, which is crucial for improving the long-term performance and reliability of the wind farm. Optimal spacing is not just a matter of simple calculations; it involves intricate modeling and simulation.

Furthermore, advanced turbine designs can reduce wake impact. For example, some turbines incorporate features designed to minimize turbulence in their wakes. This can involve specialized blade designs or modifications to the nacelle, the housing that holds the turbine's main components. These design improvements can significantly decrease the energy loss due to wake effects. Such innovations highlight the continuing evolution of wind turbine technology and its impact on efficiency.

However, implementing these strategies requires thorough understanding and careful planning. Optimizing turbine placement and controlling wake effects involves sophisticated analysis, advanced computational tools, and expert knowledge. This adds to the complexity of wind farm development, demanding precise design and precise implementation. Each choice impacts the overall efficiency of the entire farm, demanding that attention be paid to both micro\item and macro-level considerations.

Ultimately, effective wake management is a crucial element of maximizing wind farm efficiency. By integrating sophisticated modeling techniques, innovative control strategies, and advanced turbine designs, developers can significantly reduce wake losses and improve the overall energy output and economic viability of wind energy projects. Understanding the interplay of these factors ensures the creation of well-designed and highly efficient wind farms. The integration of all these factors requires both technological advancement and expert application.

5.3 Environmental Impact

The environmental impact of wind energy projects is a complex issue demanding careful consideration. While wind power is undeniably a cleaner alternative to fossil fuels, its deployment isn't without ecological consequences. Careful site selection, for example, is paramount to minimizing disruption to wildlife habitats. Birds and bats, in particular, can be vulnerable to collisions with turbine blades, a significant concern that necessitates comprehensive environmental impact assessments (EIAs) before any project commences. These assessments, rigorously conducted and independently verified, should account for both short-term and long-term effects, considering not just immediate habitat disturbance but also potential changes to local ecosystems over time. Mitigation strategies, identified and implemented through these EIAs, are crucial for ensuring the responsible development of wind energy.

Effective mitigation strategies often involve employing technologies such as bird and bat deterrent systems. These systems, ranging from acoustic deterrents to radar-based warning systems, aim to reduce collisions without negatively affecting turbine efficiency. The efficacy of these technologies is continually being refined, and careful monitoring is essential to evaluate their effectiveness and adapt strategies as needed. Beyond technological solutions, strategic turbine placement is also critical. Siting turbines away from crucial migratory routes or known habitats for sensitive species significantly minimizes the risk of wildlife collisions. This requires detailed ecological surveys and advanced modelling to predict potential impacts accurately.

Furthermore, the visual impact of wind farms is a subject of ongoing discussion. The perceived aesthetic intrusion of large turbines can lead to public opposition, underscoring the need for thoughtful integration into the landscape. This requires collaboration between developers, local communities, and environmental agencies to navigate potential conflicts and find acceptable solutions. Careful consideration of turbine design, color schemes, and overall visual integration can significantly lessen the perceived impact, mitigating public concerns while advancing wind power development. Public perception plays a significant role in project acceptance and the overall success of wind power initiatives.

Beyond direct wildlife impacts and visual concerns, other environmental aspects need careful evaluation. Noise pollution is a frequent concern, particularly for communities located close to wind farms. However, advances in turbine design are steadily reducing noise levels, and effective noise modelling techniques allow developers to accurately predict sound propagation and implement mitigating measures. These measures can range from employing quieter turbine designs to strategically placing noise barriers, ensuring compliance with environmental noise regulations, and mitigating any potential disruption to nearby communities. Careful management of electromagnetic fields is also vital, especially given increasing concerns about their potential impact on human health and the environment.

Construction of wind farms itself presents a temporary but significant environmental footprint. This includes habitat disturbance, erosion potential, and the emission of greenhouse gases during the construction phase. Mitigation requires responsible site preparation and construction practices, such as minimizing soil disturbance, employing erosion control techniques, and using low-emission equipment. Sustainable sourcing of materials and responsible waste management further contribute to reducing the overall environmental footprint of wind farm construction. The entire lifecycle, from manufacturing to decommissioning, needs this same level of thoughtful consideration. Careful planning for the decommissioning and recycling of turbine components is also crucial to minimize long-term environmental consequences. These aspects ensure that wind energy contributes positively to the environment both during its operation and its eventual removal.

Responsible development of wind energy therefore necessitates a holistic approach, where careful site selection, advanced mitigation technologies, comprehensive environmental impact assessments, and community engagement combine to minimize the environmental footprint while harnessing the benefits of clean energy. This approach ensures that the transition to renewable energy sources is environmentally responsible and sustainable, contributing to a healthier planet for future generations. The future of wind power depends on a continuous commitment to environmental stewardship, integrating technological innovation with responsible land management practices. Only through a well-balanced and considered approach can we truly unlock the full potential of wind energy while upholding our commitment to environmental protection. Transparency and continuous monitoring, along with adaptive management strategies, ensure that the environmental impact of wind energy remains under close scrutiny and that necessary adjustments are made as our understanding evolves.

6 Economic & Policy Aspects

The economic viability of wind energy projects hinges on several key factors. Initial capital costs, including turbine procurement, land acquisition, and grid connection infrastructure, represent a significant upfront investment. These costs vary greatly depending on turbine size, technology, location, and the complexity of the project. For example, offshore wind farms typically require substantially higher capital expenditures than onshore projects due to the harsher environmental conditions and increased logistical challenges. Ongoing operational expenses include maintenance, repairs, insurance, and personnel costs, all impacting the project's overall profitability. Accurate cost estimation is critical for securing financing and attracting investors, requiring detailed financial modeling and sensitivity analyses to account for uncertainties like fluctuating energy prices and potential regulatory changes. A thorough understanding of these costs, and how they differ across various scenarios, is crucial for successful project development. Government policies play a pivotal role in shaping the economic landscape of the wind energy sector. Subsidies, tax credits, and renewable portfolio standards (RPS) are common policy mechanisms used to stimulate wind energy development. These incentives reduce the financial burden on developers, making projects more attractive to investors. However, the availability and level of these incentives often fluctuate, creating uncertainty in the market. Furthermore, the effectiveness of these policies varies across different regions and jurisdictions, influencing the geographical distribution of wind energy projects. Understanding the specific policy frameworks at the national and regional levels is crucial for developers and investors to make informed decisions about project feasibility and profitability, and therefore, a careful study of relevant legislation and regulations is fundamental.

Market trends reveal a dynamic environment with significant growth potential, but also inherent risks. The global demand for renewable energy is increasing steadily, driven by concerns about climate change and energy security. This increasing demand translates to a growing market for wind turbines and related services. However, this market is subject to significant competition, both from other renewable energy sources (solar, hydro) and from conventional fossil fuels. Fluctuations in energy prices, technological advancements, and evolving government policies all impact the competitiveness of wind energy in the broader energy market. Technological improvements in turbine design and manufacturing are leading to increased efficiency and reduced costs, making wind energy more competitive. Careful market analysis, incorporating macroeconomic factors and technological forecasts, is essential for effective strategic planning and risk management in the wind energy sector. The constant evolution of this landscape mandates continuous monitoring of market signals and adapting business strategies accordingly.

Analyzing the lifecycle cost of wind energy projects provides a comprehensive view of their economic performance. This assessment extends beyond the initial capital expenditure to encompass all costs incurred over the project's lifetime, including operation, maintenance, decommissioning, and potential environmental remediation. This detailed analysis is vital for determining the overall cost-effectiveness of wind energy compared to other generation technologies, and for optimizing project design and operations to minimize long-term costs. Lifecycle cost assessments also help in evaluating the financial risks associated with different project scenarios and making informed decisions about project financing. Further refinement of these assessments considers potential future cost reductions due to technological advancements and economies of scale. A clear and well-supported lifecycle cost analysis is instrumental in securing investment and demonstrating the long-term financial viability of wind energy projects.

The interaction between economic factors and policy decisions creates a complex interplay that influences the development and growth of the wind energy industry. Policy incentives are often designed to overcome the initial higher cost of wind energy compared to fossil fuels. However, the effectiveness of these incentives can be undermined by other economic factors, such as the availability of financing, the price of raw materials, and the overall macroeconomic environment. Furthermore, policy changes can introduce uncertainty into the market, affecting investment decisions and project timelines. A comprehensive understanding of this dynamic interplay requires considering not only the explicit policy instruments but also the broader economic context, and the ability to assess the potential impact of these combined forces is vital for long-term planning and success in the field. Successful navigation of this complex terrain requires a clear understanding of policy levers and their interaction with market forces, creating a strategic framework that considers all these crucial factors.

6.1 Cost Analysis

The financial viability of any wind energy project hinges on a comprehensive cost analysis. This involves meticulous assessment across all project phases, from initial site surveys and permitting to construction, operation, and eventual decommissioning. Ignoring even minor expenses can severely skew projections and lead to unexpected financial burdens, jeopardizing the project's overall success. Therefore, a robust methodology is paramount.

Detailed cost breakdowns should encompass land acquisition or lease agreements; the purchase and transportation of turbines, foundations, and other necessary equipment; the construction of access roads, electrical infrastructure, and any required substations; and the operational and maintenance costs for the lifetime of the project. These costs are significantly influenced by factors such as the turbine's capacity, the chosen site's characteristics (terrain, accessibility, grid connection distance), and prevailing market prices for materials and labor. Careful consideration should also be given to potential delays and unforeseen expenses. Consistently monitoring these costs is critical for successful budget management.

One often underestimated cost is that of permitting and regulatory compliance. Navigating environmental impact assessments, securing necessary permits from local, state, and potentially federal authorities, and complying with various regulations can prove unexpectedly expensive and time-consuming. These processes necessitate engaging specialized consultants, lawyers, and other professionals, which adds substantially to the overall cost. Understanding and adequately budgeting for these regulatory hurdles is crucial, particularly given the considerable variation in regulatory frameworks across different geographical locations. Failing to do so can cause substantial delays and ultimately damage project feasibility.

Furthermore, the long-term operational and maintenance (O&M) costs represent a significant portion of the overall expenditure. These include scheduled maintenance, repairs, insurance, and personnel costs. Predicting these costs accurately requires considering factors such as the turbine technology's reliability, the harshness of the operating environment, and the availability of skilled technicians for repair and maintenance. A thorough assessment of the lifetime O&M costs is crucial to developing a realistic financial model for the project, often necessitating detailed simulations and risk analysis. Remember, underestimating these costs is a frequent cause of project failure.

Capital costs associated with wind energy projects are substantial. This includes the upfront investment in acquiring the necessary land, designing and engineering the wind farm, purchasing and installing the turbines, and connecting the farm to the electrical grid. Borrowing often constitutes a significant portion of the capital required, incurring interest expenses that must be factored into the financial projections. It's important to secure financing from reliable sources, perhaps involving multiple lenders and equity investors, ensuring a favorable interest rate and repayment terms. Securing funding is a critical stage, demanding a thorough and persuasive business plan supported by a meticulous cost analysis.

The environmental impact assessment is another important area. It's not just a regulatory necessity; it significantly influences project costs. The extent of the environmental study required is directly linked to the project's location and scale, potentially involving detailed ecological surveys, habitat assessments, and mitigation plans to address potential impacts on wildlife, water resources, or cultural heritage sites. These studies involve specialized consultants, adding to the overall cost. Moreover, mitigation measures might be necessary, such as adjusting turbine placement to minimize bird mortality or implementing noise reduction technologies, further increasing the expenses.

Finally, decommissioning costs should not be overlooked. This involves the dismantling of the wind turbines, the removal of foundations and other infrastructure, and the restoration of the site to its pre-construction condition. Adequate provisions for these expenses must be incorporated into the project's financial planning, preferably into a dedicated decommissioning fund. Underestimating these costs can leave developers facing substantial liabilities at the end of the project's lifespan. Accurate budgeting for decommissioning ensures long-term financial responsibility and environmental stewardship.

6.2 Policy Incentives

Government policies play a crucial role in shaping the growth trajectory of the wind energy sector. Incentives, both direct and indirect, are instrumental in driving investment, deployment, and technological innovation. Understanding these policy mechanisms is vital for stakeholders involved in developing and promoting wind power.

Subsidies, in the form of tax credits, grants, or direct payments, significantly lower the upfront costs associated with wind farm development. These financial injections are particularly crucial for projects in early stages, when investment risks are higher and securing private financing can be challenging. They help to bridge the gap between the cost of renewable energy and traditional fossil fuel sources, making wind power a more competitive option. The design of these subsidies is critical; effectively targeted programs that incentivize innovation and environmental responsibility yield the highest returns.

Production tax credits, for instance, provide a per-kilowatt-hour payment for the electricity generated by a wind farm. This mechanism directly rewards successful energy production, stimulating the operational efficiency of wind farms and reducing the long-term cost of electricity. Conversely, poorly designed subsidies can lead to inefficient resource allocation or environmental concerns. A poorly designed program might unintentionally incentivize the development of projects in locations where the wind resource is weak, leading to lower overall energy output and a less efficient use of public funds. Careful consideration of geographical factors and environmental impact assessments are paramount.

Renewable portfolio standards (RPS) represent another important policy tool. These regulations mandate that a certain percentage of a region's electricity must come from renewable sources, including wind power. This creates a predictable market demand, encouraging long-term investment and reducing uncertainty for developers. The stringency of RPS targets, along with the inclusion of various renewable sources, can influence the composition of the renewable energy mix within a specific region. For example, a more ambitious RPS targeting a higher percentage of renewable energy may accelerate the adoption of wind energy, and a broader RPS that includes other sources will offer a mix of renewable energies, leading to a more resilient and diversified energy portfolio.

Feed-in tariffs (FITs) guarantee a fixed price for the electricity generated by renewable sources over a specified period. This provides long-term price certainty and reduces the risk associated with fluctuating electricity prices. FITs can be particularly beneficial for smaller-scale wind projects or those in locations with volatile energy markets. The length of the contract and the guaranteed price directly affect the attractiveness of these policies. A longer-term contract with a competitive price would stimulate higher investment, especially considering the longer lifespan of turbines and the inherent long-term investment associated with renewable energy projects.

Beyond direct financial incentives, indirect policy support is equally critical. Streamlined permitting processes, clear regulatory frameworks, and grid interconnection policies are crucial for accelerating wind farm development. Lengthy and complicated permitting procedures can significantly delay project completion, increasing costs and hindering overall development. The establishment of clear guidelines for land use, environmental impact assessments, and grid connection reduces uncertainty and facilitates faster project development. Efficient grid interconnection infrastructure is essential for the effective integration of wind power into existing electricity grids.

Furthermore, investment in research and development (R&D) through government funding or tax breaks directly stimulates technological innovation in the wind energy sector. This can lead to improvements in turbine design, energy storage solutions, and other technological advancements that reduce costs, enhance efficiency, and increase overall competitiveness. Supporting R&D activities also contributes to the long-term sustainability of the industry, enabling it to adapt to changing market conditions.

In conclusion, effective policy incentives are essential for fostering the growth and sustainability of the wind energy sector. A well-designed policy framework combines direct financial support such as subsidies and tax credits with indirect support like streamlined permitting and R&D investment, creating a favorable environment for development and innovation. Careful consideration of policy design, geographical factors, and environmental impacts ensures that incentives are effective and sustainable in the long term. A balanced approach encompassing all these elements leads to a thriving wind energy industry, driving economic growth and advancing the transition to a cleaner, more sustainable energy future.

6.3 Market Trends

The wind energy market is experiencing a period of rapid transformation, driven by a confluence of factors. Governmental policies promoting renewable energy sources, coupled with increasing concerns about climate change and the need for energy independence, are fueling substantial investment in wind power projects globally. This surge in demand has led to intense competition among manufacturers, driving innovation in turbine design and efficiency, and pushing down costs per kilowatt-hour generated. However, this growth is not uniform across all regions.

Policy support plays a crucial role in shaping market dynamics. Countries with robust feed-in tariffs, tax credits, and renewable portfolio standards witness faster adoption rates. Conversely, regions lacking clear policy frameworks or facing regulatory hurdles may experience slower growth, highlighting the essential connection between government support and market expansion. For instance, the substantial growth in offshore wind energy projects in Europe, particularly in the UK and Denmark, is directly linked to targeted policy incentives that incentivize investment and construction. The same is not always true globally.

Technological advancements are another key driver. The continuous improvement in turbine design, including the development of larger rotor diameters and more efficient generators, has resulted in significant cost reductions. The development of smart grid integration systems also allows for more efficient management and integration of the power produced by wind farms in national grids. Simultaneously, the cost of manufacturing materials used in turbine construction is continually declining with greater economies of scale and improved manufacturing processes. Advanced materials research is further leading to the development of lighter, stronger, and more durable blades, extending their operational lifespan.

The market is also becoming increasingly diversified. While onshore wind remains a dominant sector, offshore wind is gaining momentum, driven by the prospect of higher capacity factors and access to consistently stronger winds. This shift, however, presents unique challenges related to installation, maintenance, and the cost of connecting offshore farms to onshore grids. The emergence of hybrid wind-solar projects, integrating wind and solar technologies to complement each other's energy generation profiles, further exemplifies market diversification strategies, creating resilient power systems that can effectively accommodate varying energy demands.

Analyzing market trends involves considering several key indicators. Installed capacity figures provide a snapshot of the cumulative growth of wind power. This data offers valuable insights into the expansion rate and geographical distribution of wind energy infrastructure. The cost per kilowatt-hour generated is a crucial metric, reflecting the economic competitiveness of wind power against other energy sources. Trends in this cost indicate progress in lowering the overall price of electricity produced by wind. Additionally, studying investment patterns, including venture capital inflows and public sector spending, illuminates the direction of future expansion and where investments will likely flow. The development of standardized certifications and performance testing for wind turbines and their components also impacts market trends by ensuring quality and reliability, facilitating a sense of trust.

Competitive dynamics significantly influence market developments. A global landscape of wind turbine manufacturers is vying for market share, leading to increased competition and price reductions. This environment of strong competition, while positive for consumers and the adoption of wind energy, also results in heightened pressure for manufacturers to improve efficiency and reduce costs. Mergers and acquisitions within the industry are also transforming the market structure, creating larger entities with greater resources and global reach. This consolidation can streamline production and development processes.

Finally, understanding future market projections requires careful consideration of various factors. Forecasting growth requires analyzing ongoing technological advancements, anticipating policy changes, assessing economic conditions, and considering the broader energy landscape. Reliable projections are key for guiding investment decisions, optimizing project planning, and ensuring the continued expansion of wind energy's contribution to global sustainability goals. Such forecasts depend on the ongoing developments in policy changes and related governmental incentive programs that can make the sector more appealing for investors, further expanding the adoption rates. The continuous development and exploration of new technologies is paramount for the future of wind energy, especially the advancements and improvements in turbines' efficiency and storage solutions.

7 Future of Wind Energy

Offshore wind farms represent a significant expansion of wind energy's potential. Their location, often miles from shore, allows for consistent, higher-velocity winds, resulting in significantly increased energy generation compared to their onshore counterparts. However, the challenges are considerable; construction and maintenance are more complex and costly, requiring specialized vessels and equipment. The environmental impact, particularly on marine life, requires meticulous planning and mitigation strategies. Further research into innovative foundation designs, such as floating platforms for deeper waters, is crucial for unlocking the full potential of this vast resource, and for making it economically competitive with more established energy sources. Successfully navigating these hurdles promises a substantial contribution to global energy sustainability.

Hybrid systems, combining wind energy with other renewable sources like solar power, offer a compelling pathway to enhance grid stability and reliability. By integrating complementary energy sources, we can address the intermittency inherent in wind power. When the wind isn't blowing, solar panels can provide power, ensuring a consistent energy supply and significantly reducing reliance on fossil fuels. Clever energy management systems, incorporating advanced forecasting and storage technologies, are crucial for maximizing efficiency and minimizing waste within these hybrid configurations. The synergy between these technologies will be key to reaching wider adoption and ensuring a seamless transition to cleaner energy sources, with the potential to radically reshape our energy grids in a climate-conscious manner.

Technological advancements continue at a rapid pace, driving significant improvements in wind turbine design and efficiency. We are witnessing the development of larger and more efficient blades with innovative aerodynamic profiles, maximizing energy capture from even light breezes. The integration of advanced materials, such as lighter yet stronger composites, leads to longer turbine lifespans and reduced maintenance costs. Moreover, the deployment of more sophisticated control systems enhances energy capture and optimizes grid integration, and this constant innovation contributes to the ongoing reduction in the cost of wind energy, creating the market conditions necessary to drive global growth. This ongoing research and development, alongside the constant push for sustainability and long-term viability, is paramount to the ongoing success and improvement of wind-based renewable energy technologies.

Sustainability considerations are integral to the future of wind energy. Minimizing the environmental impact of wind farm construction and operation is paramount. Careful site selection, avoiding environmentally sensitive areas, and implementing effective noise and visual impact mitigation strategies are essential. The lifecycle assessment of turbines, considering material sourcing, manufacturing processes, and eventual decommissioning, is crucial for responsible development. Addressing these concerns ensures that wind energy contributes to a truly sustainable future, not just in relation to its output, but in relation to its entire contribution to the environment. Transparency and responsible governance are key.

Energy storage is a crucial component for the full realization of wind energy's potential. The intermittent nature of wind necessitates effective energy storage solutions to smooth the supply and ensure consistent electricity. Emerging battery technologies, including improved lithium-ion batteries and emerging alternatives such as flow batteries, offer promising solutions, while pumped hydro storage continues to be a cost-effective solution in specific geographic locations. Significant investment and research and development efforts are needed to improve the efficiency, reduce the cost and improve the lifecycle management of these storage solutions, ensuring that wind energy can deliver a dependable energy supply. This will remain a key part of the ongoing discussion surrounding wind energy as an economically viable and sustainable solution.

7.1 Offshore Wind Farms

Offshore wind farms represent a significant leap forward in harnessing wind energy's potential. Their location, often far from land, allows for consistent, high-velocity winds, resulting in significantly increased energy generation compared to onshore facilities. This increased capacity is crucial in meeting growing global energy demands while reducing reliance on fossil fuels. However, establishing these facilities presents unique engineering and logistical challenges that necessitate innovative solutions.

The sheer scale of offshore wind farms demands robust infrastructure. Subsea cables, capable of withstanding the harsh marine environment and transmitting massive amounts of power, are essential. Their design and installation require specialized vessels and techniques, pushing the boundaries of subsea engineering. Regular maintenance and repair of these underwater systems are equally critical, necessitating sophisticated remotely operated vehicles (ROVs) and underwater inspection methods. The logistical complexities involved in transporting materials and personnel to offshore sites are also considerable, demanding efficient and cost-effective solutions, perhaps through the use of advanced floating platforms or purpose built vessels.

Beyond the infrastructure, the environmental impact requires careful consideration. Careful planning and mitigation strategies are vital to minimize disruption to marine ecosystems. Environmental impact assessments (EIAs) are not just a regulatory requirement; they are essential tools for understanding and mitigating potential risks to marine wildlife, including birds, fish, and marine mammals. These assessments should involve extensive studies of marine life, including noise pollution assessments that account for potential impacts on sea mammals and the ecosystem's delicate balance. Adaptive management strategies, incorporating real-time monitoring and data analysis, can allow for prompt adjustments to operational practices, minimizing negative impacts throughout the project's lifecycle. This proactive approach is paramount for responsible and sustainable offshore wind farm development.

Turbine design for offshore environments also presents distinct challenges. The turbines must withstand extreme weather conditions, including high winds, large waves, and saltwater corrosion. This requires the use of specialized materials and robust designs capable of enduring these extreme conditions. Furthermore, the installation and maintenance of these massive turbines pose significant logistical challenges, demanding the use of specialized heavy-lift vessels and highly skilled personnel. The accessibility of offshore facilities directly impacts maintenance costs and schedules, emphasizing the need for high reliability and robust design from the start.

The economic viability of offshore wind farms hinges on various factors, including technological advancements, government policies, and market conditions. Technological improvements, leading to more efficient and durable turbines, are crucial for reducing capital costs and improving energy output. Substantial government support, through subsidies, tax credits, and regulatory frameworks, is vital to incentivize investment and encourage the growth of this crucial sector. Moreover, stable and predictable energy markets are essential to ensure long-term profitability and attract investor interest. This interplay between technology, policy and market dynamics underscores the importance of a collaborative approach, bringing together researchers, policymakers, and the private sector to create a sustainable pathway for offshore wind energy development.

Finally, the integration of offshore wind farms into existing power grids presents a significant hurdle. The transmission of electricity from these remote locations necessitates the development of high-voltage direct current (HVDC) transmission systems that can efficiently transport large quantities of electricity over long distances. The integration of these vast energy sources demands meticulous coordination and sophisticated grid management systems, capable of handling the intermittent nature of wind power and ensuring grid stability. This complex interplay between offshore wind farms and the power grids highlights the critical need for investment in smart grid technology and advanced grid control strategies, ensuring the reliable delivery of renewable energy to consumers. The successful integration of offshore wind farms into the energy mix underscores the essential role of technological innovation, policy support and smart grid management in achieving sustainable energy goals.

7.2 Hybrid Systems

Harnessing the full potential of wind energy often necessitates a multifaceted approach. Integrating wind power with other renewable sources, such as solar or hydropower, creates a hybrid system, offering enhanced reliability and efficiency. This synergistic approach mitigates the inherent intermittency of wind, providing a more consistent and predictable energy supply. By combining different renewable energy sources, we can create a more resilient and sustainable energy infrastructure.

Consider a scenario where a wind farm is situated alongside a solar array. During periods of high solar irradiance, the solar panels generate substantial power, supplementing the wind turbines' output. Conversely, during periods of low solar radiation, or at night, the wind turbines take over, ensuring a continuous flow of electricity. This complementary relationship minimizes fluctuations in energy production, leading to increased grid stability and reduced reliance on fossil fuel backup systems. This is a highly efficient and beneficial integration that leads to an improved energy supply.

Careful planning and design are essential for effective hybrid system integration. Factors such as geographical location, resource availability, and energy demand must all be carefully considered. A thorough site assessment is required to determine the optimal placement of wind turbines and solar panels, maximizing energy capture while minimizing environmental impact. Sophisticated energy management systems are needed to efficiently integrate the different sources, optimizing energy flow and distribution. Detailed site analysis includes detailed meteorological studies, geographical constraints, environmental considerations, and detailed simulations to optimize for maximum energy production.

Advanced power electronics play a vital role in seamlessly integrating diverse energy sources. Intelligent inverters and grid-connected systems manage power flows between different generators, ensuring smooth and efficient transitions between wind and solar power. They also help regulate voltage and frequency, contributing to grid stability. Sophisticated control algorithms ensure that the hybrid system operates optimally, adapting to changing environmental conditions and fluctuating energy demand. Careful consideration of grid stability requirements is paramount to successful implementation.

The economic benefits of hybrid systems are compelling. The reduced reliance on backup power generation translates to substantial cost savings. Furthermore, hybrid systems can qualify for various incentives and subsidies offered to renewable energy projects, further enhancing their economic viability. In essence, combining various renewable resources results in significantly reduced long-term operating expenses, increased potential for profits, and a more environmentally friendly power solution. We must remember the long-term sustainability and cost-effectiveness factors of such implementations.

Moreover, hybrid systems can significantly enhance the environmental sustainability of energy production. By reducing reliance on fossil fuels, they contribute to lower greenhouse gas emissions and a smaller carbon footprint. This ecological benefit is amplified when the selected renewable energy sources have minimal environmental impact, furthering the positive contribution to a cleaner energy future. Moreover, efficient hybrid systems drastically improve the environmental friendliness of power generation and minimize the negative impact on the environment.

However, challenges remain. The initial investment for a hybrid system may be higher than for a single renewable energy source project. The need for sophisticated energy management systems also adds to the complexity. Overcoming these challenges requires careful planning, technological advancements, and supportive policies. However, the long-term benefits outweigh the initial costs and complexities, offering a sustainable and economically viable path to a cleaner energy future. Careful consideration of these complexities and strategic planning ensures long-term success and reduced operational problems.

Ultimately, the development and implementation of efficient hybrid systems represents a crucial step towards achieving a sustainable energy future. By combining different renewable sources, we can create a more resilient, efficient, and environmentally friendly energy infrastructure. The benefits extend beyond the individual projects, contributing to broader goals of reducing carbon emissions and enhancing energy security. Investing in this approach is not just an environmentally responsible decision but also a strategically sound one that ensures energy self-sufficiency and resilience. Through collaborative innovation and strategic resource allocation, the future of energy is brightly lit by such integrated approaches.

7.3 Technological Advancements

Advances in materials science are revolutionizing wind turbine design. Lightweight yet incredibly strong composites, like carbon fiber reinforced polymers, are replacing traditional materials, leading to larger blades capable of capturing more wind energy. These advanced materials also enhance durability, reducing maintenance needs and extending the operational lifespan of turbines significantly. This translates to lower lifecycle costs and a more sustainable energy solution. Furthermore, research into self-healing materials promises to further minimize downtime and repair costs, making wind power even more economically attractive. The development of these materials isn't simply an incremental improvement; it represents a fundamental shift in the capabilities and longevity of wind turbines. It's a crucial step toward widespread adoption of this renewable energy source.

The integration of artificial intelligence and machine learning is transforming wind energy operations. AI algorithms can analyze vast amounts of data from various sources, including weather forecasts, turbine sensor readings, and grid demand, to optimize turbine performance in real-time. Predictive maintenance models, powered by machine learning, allow for proactive identification of potential equipment failures, minimizing costly downtime and enhancing operational efficiency. This data-driven approach surpasses traditional methods, enabling more precise control over energy output and maximizing energy yield. The deployment of AI isn't merely about automation; it's about achieving an unprecedented level of optimization and efficiency in wind farm management. This technology will greatly enhance the economic viability of wind energy projects and expedite their development.

Beyond the turbine itself, advancements in energy storage are crucial for addressing the intermittency of wind power. While wind doesn't always blow consistently, effective energy storage solutions can buffer fluctuations in energy supply. Battery technology is rapidly advancing, with improvements in energy density, charging speeds, and lifespan. Furthermore, innovative storage approaches, such as pumped hydro storage and compressed air energy storage, offer alternative solutions for larger-scale energy management. These advancements allow wind farms to deliver a more consistent and reliable power supply, contributing to grid stability and improving the integration of renewable energy sources into existing infrastructure. Effective energy storage isn't simply a technological challenge; it is the key to unlocking the full potential of intermittent renewable energies such as wind power.

Another area of significant progress lies in the development of innovative control systems for wind turbines. Advanced control algorithms are designed to maximize energy capture under varying wind conditions. These algorithms optimize blade pitch angle, yaw control, and generator speed, constantly adapting to changes in wind speed and direction. The integration of sophisticated sensors and actuators allows for precise control, minimizing energy losses and enhancing overall efficiency. Furthermore, these control systems can incorporate real-time data from weather forecasts to predict upcoming wind conditions and proactively adjust turbine operations for optimal performance. The complexity of modern wind turbine control systems is impressive, making these machines not just power generators but advanced, self-regulating systems.

Offshore wind farms are increasingly important due to their access to consistent and stronger winds. However, the harsh marine environment presents unique challenges, requiring specialized turbine designs and robust construction techniques. Floating offshore wind turbines, for example, allow for deployment in deeper waters where fixed-bottom turbines are impractical. Advancements in materials science, such as corrosion-resistant coatings and stronger mooring systems, are enabling the development of more reliable and cost-effective offshore wind farms. These projects represent a significant expansion of wind energy's potential, pushing the boundaries of engineering and technology. The future of wind power significantly depends on success in offshore development, opening up vast untapped energy resources.

Finally, the ongoing research into hybrid renewable energy systems is significantly enhancing wind energy's role in the energy transition. Hybrid systems combine wind power with other renewable energy sources, such as solar power or hydropower, to create a more diversified and reliable energy supply. These systems can leverage the complementary characteristics of different renewable sources, mitigating the intermittency issues associated with individual technologies. This approach leads to increased energy security and enhanced grid stability, demonstrating a more robust and sustainable energy future. The development of these sophisticated hybrid energy systems will be one of the most significant factors in making renewable power the new standard in the world's electrical grids.

7.4 Sustainability

The sustainability of wind energy is paramount, not just for environmental reasons, but for the long-term viability of the industry itself. We've explored the technological advancements that enhance efficiency and output; now, let's delve into the crucial elements that ensure wind power remains a responsible and enduring energy solution. This involves considering the entire lifecycle of a wind turbine, from material sourcing to decommissioning, and minimizing environmental impact at every stage.

Manufacturing wind turbines requires significant resources. The selection of materials is critical. Prioritizing recycled materials and those with minimal embodied energy—the energy used to create the material itself—is vital. Furthermore, exploring sustainable manufacturing processes, such as reducing water usage and minimizing waste, is crucial. The adoption of circular economy principles, focusing on reuse, repair, and recycling, will significantly reduce the environmental footprint of wind turbine production. This proactive approach isn't merely a matter of good practice; it directly impacts the overall sustainability and economic competitiveness of the wind energy sector. By being responsible in the manufacturing phase, we lay a strong foundation for a greener and more financially resilient industry. This mindful approach to materials and manufacturing should be viewed as a fundamental pillar of sustainable development within the wind energy sector.

Transportation and installation of turbines also pose environmental challenges. Optimizing logistics, employing fuel-efficient transport, and choosing installation methods that minimize ground disturbance are all essential steps. Minimizing the carbon footprint associated with the transportation phase becomes increasingly critical as the size and weight of turbine components continue to increase with technological advancements. Efficient logistics planning, such as utilizing optimized shipping routes and employing multimodal transportation, contributes significantly to environmental protection. Strategic planning during the transportation stage, minimizing the use of fossil-fuel-dependent transport and embracing renewable energy-powered alternatives, can considerably reduce the sector's carbon emissions.

Operation and maintenance of wind farms require careful consideration. Regular inspections and timely repairs not only maximize energy output but also prevent premature failures, reducing the need for replacements and minimizing waste. Moreover, the use of biodegradable lubricants and efficient maintenance strategies can further enhance the sustainability profile. Developing strategies for efficient maintenance operations and leveraging renewable sources of energy for powering maintenance vehicles further contributes to reducing the sector's environmental impact. This proactive approach minimizes disruptions and prevents unforeseen complications that could necessitate substantial repairs and replacements, ultimately contributing to the overall economic and ecological viability of wind energy operations.

Finally, decommissioning and recycling of wind turbines at the end of their lifespan present both challenges and opportunities. Developing effective recycling processes for composite materials, such as fiberglass, is critical to avoid landfill waste. The recovery and reuse of valuable metals from turbine components also offer economic benefits. Innovative approaches such as employing robotics to streamline the dismantling process and making use of recovered materials in new turbine construction can establish circularity. Considering the whole lifecycle, we can enhance the sustainability of the industry by reducing landfill waste, recovering valuable components, and establishing a recycling-oriented circular economy that promotes resource efficiency and reduces environmental impact.

Sustainable land management is also essential. Careful site selection minimizes disruption to ecosystems and wildlife habitats. Innovative approaches like co-locating wind farms with other land uses, such as agriculture or grazing, can increase biodiversity and enhance the economic value of the land. Moreover, incorporating measures to minimize noise pollution and visual impact on the landscape is crucial for public acceptance. Thoughtful planning, coupled with the utilization of land for multiple purposes, can contribute to greater economic efficiency, ecological harmony, and enhanced public support. These strategies are not simply environmental considerations, but key factors in ensuring the long-term success and societal acceptance of wind energy projects.

In conclusion, the sustainability of wind energy isn't just an environmental consideration; it's inextricably linked to the economic and social viability of this critical energy source. By addressing every stage of the wind turbine's lifecycle—from material sourcing to decommissioning—and prioritizing environmentally responsible practices, we ensure that wind power continues to play a significant role in a cleaner, more sustainable energy future. A holistic and proactive approach towards sustainability is not merely an ethical imperative but is also a strategic necessity for the lasting success and broader societal acceptance of wind energy, ensuring its position as a cornerstone of a sustainable energy future. This comprehensive approach underpins the long-term viability and positive societal impact of wind energy.

7.5 Energy Storage

The intermittent nature of wind power necessitates effective energy storage solutions. Without them, the fluctuating supply of electricity generated by wind turbines can disrupt grid stability and limit the overall contribution of wind energy to the power system. Various storage technologies are currently being developed and deployed, each with its own advantages and disadvantages. Choosing the right technology depends heavily on the scale of the wind farm, the specific grid needs, and the overall cost-effectiveness. This chapter will explore those considerations.

Pumped hydro storage, a mature technology, remains a prominent choice for large-scale energy storage. It involves pumping water uphill to a reservoir during periods of surplus wind energy, then releasing it to generate electricity when demand is high. While geographically constrained, its long lifespan and high energy density make it a competitive option for many regions. However, its significant upfront capital investment and potential environmental impacts are crucial factors to consider, and careful environmental assessments must always be carried out.

Battery storage is gaining rapid traction, particularly with the advancements in lithium-ion technology. These systems offer flexibility, shorter response times, and modular scalability, making them suitable for both large-scale and smaller, distributed applications. However, concerns remain regarding battery lifespan, recycling challenges, and the environmental footprint of their manufacturing. Ongoing research into safer, more sustainable battery chemistries, like solid-state batteries, is crucial for widespread adoption. Their declining cost and improving performance profile certainly make them a strong contender for the future.

Compressed air energy storage (CAES) offers a unique approach to energy storage, using excess wind energy to compress air into underground caverns or depleted gas fields. The compressed air is then used to drive turbines and generate electricity when needed. This technology benefits from its relatively low environmental impact and potential for large-scale deployments, however, the need for suitable geological formations limits its geographical applicability. Furthermore, energy losses during the compression and expansion processes can affect overall efficiency. Careful site selection and optimization of the entire system are paramount.

Thermal energy storage (TES) utilizes the heat generated during electricity production to store energy. This heat can then be used to generate electricity later, typically via a thermal power cycle. TES systems offer potential advantages in terms of scalability and cost-effectiveness, especially when integrated with concentrated solar power plants. Yet, challenges in heat transfer efficiency and thermal losses can limit overall performance. Different TES materials, such as molten salts and various types of rocks, each exhibit distinct performance characteristics, requiring careful selection based on the specific application.

Flywheel energy storage systems utilize rotating mechanical devices to store kinetic energy, which can then be converted into electricity upon demand. Flywheels offer rapid response times and a high cycle life, making them suitable for applications requiring fast frequency regulation. However, energy density limitations currently restrict their application to relatively smaller-scale systems. Furthermore, the mechanical components' precise engineering and maintenance requirements contribute to higher costs compared to some other alternatives. Future developments could enhance their overall efficiency and broaden their applicability.

The optimal energy storage solution depends on a multitude of factors, including the specific characteristics of the wind farm, the local grid infrastructure, and the economic considerations of each option. A comprehensive assessment involving technical feasibility, economic viability, and environmental impact is required before committing to any particular technology. The future likely involves hybrid approaches, combining different energy storage technologies to leverage their respective strengths and mitigate their weaknesses. This synergistic approach promises a more resilient and efficient integration of wind energy into the power grid, maximizing the utilization of this clean and renewable resource. Continued innovation and research are essential for further reducing the cost, enhancing the performance, and improving the environmental profile of energy storage systems, paving the way for a truly sustainable energy future.

www.ingramcontent.com/pod-product-compliance
Lightning Source LLC
LaVergne TN
LVHW022020200926
848481LV00007B/2455
* 9 7 9 8 3 4 8 4 9 4 5 9 9 *